IL    2019

# SPECTACULAR YOU

*An IVF Love Story*

AuthorHouse™
1663 Liberty Drive
Bloomington, IN 47403
www.authorhouse.com
Phone: 1 (800) 839-8640

Published by AuthorHouse 08/02/2017

ISBN: 978-1-5246-9561-3 (sc)
ISBN: 978-1-5246-9562-0 (e)
ISBN: 978-1-5246-9563-7 (hc)

Library of Congress Control Number: 2017908944

Print information available on the last page.

authorHOUSE®

# SPECTACULAR YOU

*An IVF Love Story*

## Kate Pache

*This book is dedicated to all of the Moms and Dads
who never gave up on their dreams of parenthood.*

We knew from the beginning. We knew from the start.
Before you grew in Mom's belly, you grew in our hearts.

What would you be like? What would you like to do?
Whenever we slept, we were dreaming of you.

So we waited for your arrival. But the months turned to years.
And when you still weren't here, our eyes filled with tears.

But then we met a doctor. She said, "I know what to do!"
And she told us of her plan to make spectacular YOU!

So the Doctor took Mom's eggs, while Mom took a nap.

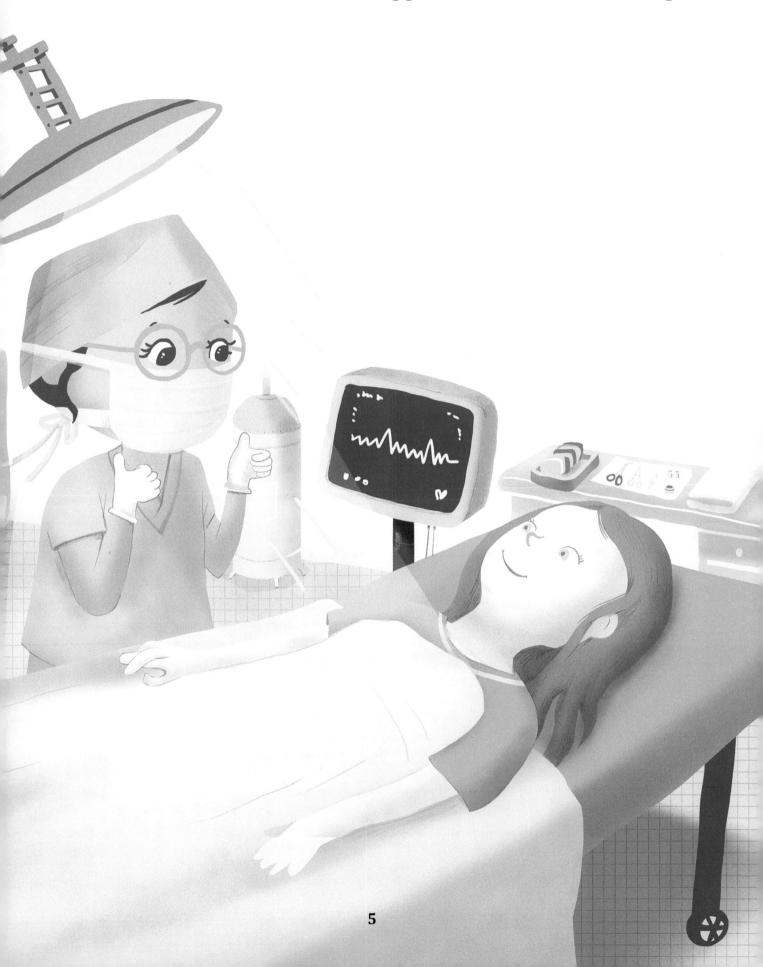

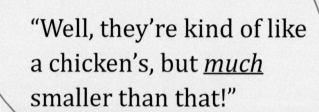

"Well, they're kind of like a chicken's, but _much_ smaller than that!"

Dad's don't have eggs but Dad's cells helped too.
Because that part of Dad made up half of you!

The scientist in the lab, on that very same day, placed one of Dad's cells into each of Mom's eggs.

After she finished, she put Mom's eggs in a dish.
She closed up the lab and made one little wish.

Then, later that night, something wonderfully strange happened there in that dish. Those eggs started to change!

ssshhh...
embryos
sleeping

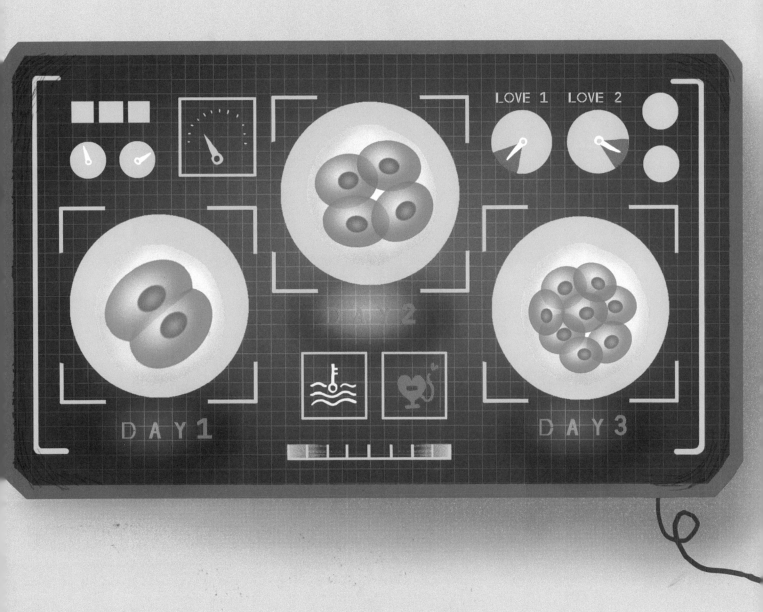

Then over 3 days, you started to grow. From 2 cells to 8 cells!
WOW! Look at you go!

THAT'S YOU!

Just 2 more days later, you grew into a ball
surrounded by water and a precious cell wall.
(There's a reason, we're sure, why they called you a BLAST.
But we think it's because you grew big so fast!)

Then Doctor said, "Mom needs a little more rest to transform her belly into a nest."

In a wee little freezer, you took a rest too,
on a bed made of snowflakes in your tiny igloo.

Just a few more weeks later
(it was all we could bear),
we exclaimed, "We're both ready!"
So, the scientist, with care, woke you up
from your slumber on your bed of snow.
Then she warmed you right up, from your
nose to your toes!

You were placed in Mom's belly while we all said a prayer that you'd continue to grow 'cause you liked it in there.

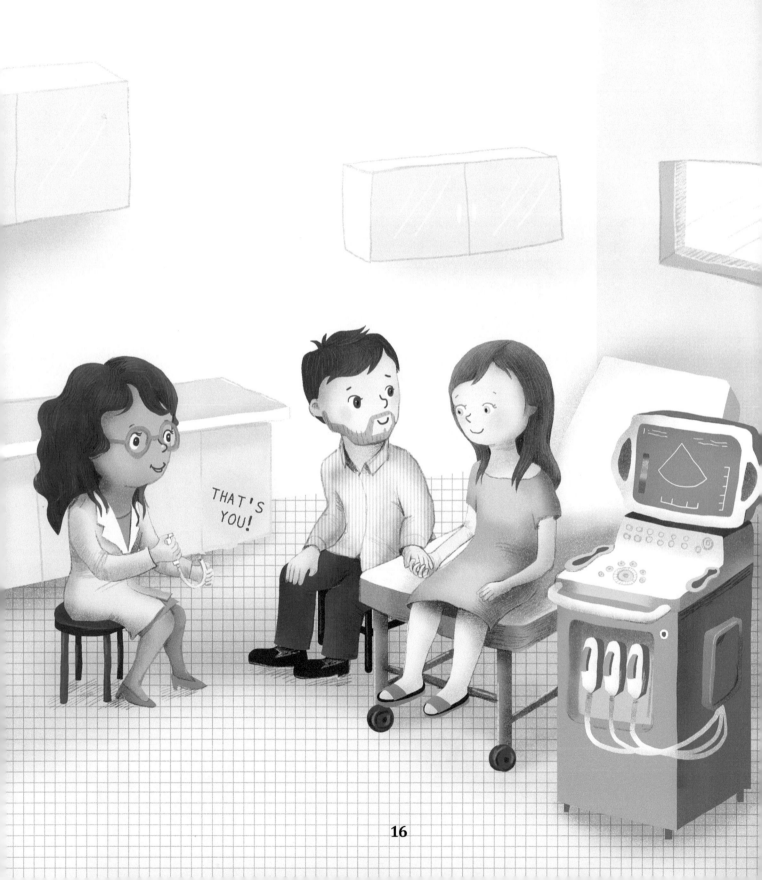

And grow you sure did! Why, you transformed Mom's figure!
We'd see every day that you kept getting bigger.

'Til that day we met you. Oh, that day that you came into both our lives. We were never the same.

And, Oh, how we LOVE YOU! Now our family is complete.
And this is your story.

(Now isn't that neat?)

The End ♡